Lilla C. Perry

From the Garden of Hellas

Translations into verse from the Greek anthology

Lilla C. Perry

From the Garden of Hellas
Translations into verse from the Greek anthology

ISBN/EAN: 9783337083236

Printed in Europe, USA, Canada, Australia, Japan

Cover: Foto ©berggeist007 / pixelio.de

More available books at **www.hansebooks.com**

FROM THE GARDEN OF HELLAS

FROM THE GARDEN OF HELLAS

TRANSLATIONS INTO VERSE FROM THE GREEK ANTHOLOGY

BY

LILLA CABOT PERRY

NEW YORK
UNITED STATES BOOK COMPANY
SUCCESSORS TO
JOHN W. LOVELL COMPANY
150 WORTH ST., COR. MISSION PLACE

JAMES RUSSELL LOWELL.

Life's brightest memories around you cling.
  Child, girl, and woman, I have loved you long.
My friend, my poet, if I too may sing,
  To you is dedicate my wreath of song.

Yet it should be of fairer flowers than grow
  Within my garden wall. This wreath I twine
In Beauty's fields, where deathless blossoms glow,
  The hand that gathers them alone is mine.

INDEX OF EPIGRAMS

*IN THE ORDER FOUND IN THE PALATINE
ANTHOLOGY.*

BOOK V.

BOOK VI.

BOOK IX.

TRANSLATOR'S PREFACE.

THERE have been many translations, in whole or in part, in verse or in prose, of that wonderful collection of the inscriptions, epigrams, and love-songs of Ancient Greece—a literature in itself known as the Greek Anthology.

In prose, however, there is lost much of the delicate fragrance and charm of these artfully devised, concise, significant poems. In verse, on the other hand, there has been admitted too much poetic license, and the simple phrases of the Greek have been too often elaborated and decorated, when they were not mistranslated or distorted, to fit the exigencies of rhyme and please a different poetic taste. It has also been the custom to select for translation only those epigrams that are most modern in sentiment, only those that would generally be deemed the most poetic. In this way one misses the infinite variety of the Anthology, its representative quality, its contrasting notes of personality, its kaleidoscopic harmony of local color.

This small book is an attempt at a selection that shall be fairly representative of all the many classes of poems, except those that, for obvious reasons, are untranslatable, and in every case the translator has aimed at literal fidelity to the original. To have imitated the Greek metres would have been a rash experiment in English, and even if successful would have been monotonous. It has therefore been deemed advisable to use, instead of metres familiar to the ancients, those familiar to ourselves.

FROM THE GARDEN OF HELLAS.

EPIGRAMS

FROM THE PALATINE ANTHOLOGY.

MELEAGER.

BOOK V., EPIGRAM 57.

That butterfly, my soul, if thou wouldst burn,
O cruel Love! too often with thy flame,
Itself has wings to fly and ne'er return.

BOOK V., EPIGRAM 136.

Pour to the health of Heliodora, pour
Again to Heliodora, and once more,
Her sweet name mingling with each cup of wine.
From flowers of yesterday a garland twine
　　Dewy with perfumes to her memory.
　　Love's roses weep that she is not with me!

1

Book V., Epigram 139.

By the god Pan of Arcady I vow
Sweet is thy singing, Zenophil, and thou
Sweetly can'st play the lyre.—Where can I flee
From all thy various charms besieging me?
Not for a moment will they let me rest.
Now 'tis thy slender form in beauty drest,
Now 'tis thy voice, thy grace.   What do I say?
It is thy*self* for whom I burn alway!

Book V., Epigram 141.

Yes, I call on Love to witness,
I would rather lend my ear
To the voice of Heliodora
Than Apollo's music hear!

Book V., Epigram 143.

On Heliodora's head the loveliest wreath
Pales by the beauties that are seen beneath.

Book V., Epigram 144.

Now the white violet once more is here
And the narcissus, lover of the rain ;
And lilies on the hills are come again.
But there's a flower of flowers to lovers dear,
More fragrant than them all, for like a rose
Her opening charms doth Zenophil disclose.
In vain, O fields ! your beauties you display,
All your gay smiling flowers you show in vain.
Who once that lovely child meets by the way
Nor heeds nor sees your wreathèd charms again.

BOOK V., EPIGRAM 147.

The violet white and laughing lilies I
Weave with the myrtle and narcissus shy,
And the sweet crocus with them do I twine,
And hyacinth as purple as rich wine,
And roses, flowers that lovers find most fair ;
So that my wreath, by Heliodora's grace,
I may upon her perfumed temples place
And crown with flowers her richly curling hair.

BOOK V., EPIGRAM 148.

Surely, while talking, Heliodora may
Surpass the Graces' selves in grace some day !

BOOK V., EPIGRAM 155.

Heliodora ! Love hath fashioned thee
    From out my very heart.
Heliodora, sweet-voiced, unto me
    As my soul's soul thou art.

4

Book V., Epigram 163.

O Bee! why touchest Heliodora's cheek?
    Feaster on flowers! why leav'st the cups of spring?
Would'st have me know that she too feels of Love
    The sweet, the unendurable, bitter sting?
Thus say'st thou, loved of lovers? Then begone!
Depart! for long thy message have I known!

Book V., Epigram 171.

The cup laughs with joy to be touched, as she sips,
    By the eloquent mouth of the fair Zenophil.
Ah, happy the cup! How I long for those lips
    That my whole heart and soul in a breath they
      may steal!

BOOK V., EPIGRAM 178.

" LOVE FOR SALE."

Who'll buy him, sleeping in his mother's arms?
　Who'll buy?　Who'll buy?
How dare I cherish him who only harms?
　That will not I.

Two wings, one snub nose, ten sharp-scratching nails.
　Who'll buy?　Who'll buy?
Sometimes he laughs, or if that naught avails,
　Why then he'll cry.

A mule, a chatterbox, sharp-eyed and wild,
A very monster is this cruel child;
E'en in his mother's pain he finds his joy.
Come!　Some out-going merchant buy the boy!
But now he pleads, he weeps,—cheer up, then.　See!
I'll sell thee not, stay with Zenophile!

BOOK V., EPIGRAM 182.

Tell her this, Dorcas! Tell her once again,
A third time, Dorcas, tell her everything.
Run, don't delay, fly! Wait a minute, wait
A moment longer, Dorcas! Whither haste
Before the whole thou know'st? Add only this
To what I said before,—but trifle not.
Say, only say—no, Dorcas, tell her all.
Why should I send you, Dorcas, for with you
I go myself? My message I precede.

Book VII., Epigram 182.

At the bride's gates the lotos flutes were sounding
   All yesterday, doors swinging to and fro.
This morn for Clearista all are weeping,
   Their song of Hymen changed to dirge of woe.

Her bridegroom, Death ; she'll have no other wedding,
   For him she looseneth her virgin zone.
The very torches for her bridal burning
   Shall light her trembling feet to Acheron.

Book VII., Epigram 195.

Cicada, you who chase away desire,
Cicada, who beguile our sleepless hours,
You song-winged muse of meadows and of flowers,
Who are the natural mimic of the lyre,
Chirp a familiar melody and sweet,
My weight of sleepless care to drive away;
Your love-beguiling tune to me now play,
Striking your prattling wings with your dear feet.
In early morning I'll bring gifts to you
Of garlic ever fresh and drops of dew.

BOOK VII., EPIGRAM 207.

ON A HARE.

From my mother's teats they tore me,
Little long-eared hare, and bore me,
   The swift-footed, from her breast.
Phanium, soft-handed, fed me
On spring flowers and nourishèd me,
   Fondling in her lap to rest.

No more for my mother sighing
Feasting daintily, then dying;
   I by too much food was slain.
And she buried me with weeping
Near her house, that she, while sleeping,
   Me in dreams might see again.

BOOK VII., EPIGRAM 461.

Lightly, O Mother Earth! on Œsigenes rest:
Lightly his foot on thee was ever pressed.

BOOK VII., EPIGRAM 476.

TO HELIODORA.

Heliodora, tears that pierce the earth,

The last gift of my love, receive from me

Beyond the grave; tears shed most bitterly!

Alas! Upon thy tomb there is no dearth

Of tears, that in past joy have had their birth,

Poured in libation to the memory

Of faithful love, thus consecrate to thee,

To thee, though dead, my only thing of worth.

Where is my flower that Hades plucked? Ah!
    where?

An idle sacrifice to Acheron!

Dust now defiles its petals blooming fair,

Hades hath stolen her, hath stolen her!

All-mother Earth, I pray thee, gently bear

Upon thy breast one whom all weep, now gone!

BOOK IX., EPIGRAM 363.

TO SPRING.

The gusty winter from the sky now clears,
    Spring with her rosy hours
    Comes bringing smiles and flowers,
And sombre earth in fresh young green appears,

And all the budding plants new leaves adorn,
    The roses open, while
    The fields greet with a smile
The tender dew that's brought them by the morn.

The shepherd, joyous on the mountain height,
    Upon his pipe now plays
    His songs and roundelays,
And goatherds in their snowy kids delight.

The harmless zephyr fills the swelling sail,
    The wind the sailor craves
    Now sweeps o'er the broad waves,
While on the shore is heard a distant "Hail!"

" Hail, Dionysus, patron of the grape ! "
    From those whose temples twine
    The blossoms of the vine.
Then from the shaggy herd the bees escape,

And cluster on the beehives, fashioning
    Their white and beauteous cells,
    From which fresh honey wells.
Around the house are swallows twittering,

And clear-voiced birds are singing everywhere.
    Along the river side,
    Where swans sedately glide,
And kingfishers are darting through the air.

At dark the nightingale no more is mute.
    Since bees and birds find voice,
    And thick-fleeced sheep rejoice
To hear the shepherd pipe upon his flute,

Since now the Bacchic choruses outring,
    And leaves and plants are glad,
    And many a sailor lad,
Should not I too sing beauty in the spring?

BOOK XII., EPIGRAM 47.

Yet in his mother's lap, at break of day,
A baby throwing dice, Love played my soul away.

BOOK XVI., EPIGRAM 134.
(Appendix Planud.)

ON A STATUE OF NIOBE.

O Niobe! thou child of Tantalus!
   Give ear thou must to me,
A messenger of grief; most piteous
   My tidings unto thee!

For Phœbus' fatal arrows did'st thou bear
   Thy noble sons, ah me!
Unloose, unloose the band that binds thy hair!
   Alas! e'en more I see.

For on thy daughters who to thee have fled
  The wave of death o'erflows :
This one across thy knees is lying dead,
  And that one crouching knows

Not where she may escape the avenging dart.
  One from still living eye
Turns a last look at thee, while on thy heart
  One but finds leave to die.

And thou, the mother, who thy speech's flood
  Checked not and must atone,
With grief and horror all thy flesh and blood
  Is frozen into stone.

BOOK XVI., EPIGRAM 213.
(Appendix Planud.)

I'll flee thee, Love, to my last breath,
  Though swift thy wings, though sharp thy ar-
    rows.
Yet what avails it ?   Even Death
  Knows thee, since Pluto felt thy sorrows.

ANTIPATER.

Not of Themistocles * am I the tomb;
No! A Magnesian monument I am
To the ungrateful rancor of the Greeks.

Albeit I am dead, this cruel sea
    Disturbs me, Lysis, buried 'neath a rock,
    Breaking upon my silent tomb with shock
Of heavy, booming waves pursuing me.
Why did you place me near this ocean? Why?
    For not in stately pleasure-ships of cost,
    But in a humble merchant's boat I tossed,
And where I sought my living did but die.

* Themistocles died in exile at Magnesia.

Book VII., Epigram 464.

Aretemias, when, from the infernal bark
Thy foot thou placedst on Cocytus' shore,
Bearing in thy young arms thy new-born babe,
The lovely Dorian girls, all pitiful
At hearing of thy fate, would question thee.
And then through tears thou utter'dst these sad
     words :
" Twin children have I brought into the world ;
One with my husband, Euphron, did I leave,
This other I bring with me to the dead."

BOOK VII., EPIGRAM 713.

Few were thy words, Erinna, few thy songs,
　　And yet in them the Muses found delight!
We lesser singers, in unnumbered throngs,
　　Perish and are forgotten ; but the night
Of black oblivion, with shadowy wing,
　　Ne'er sweeps thy gentle image from our sight.
Sweeter the swan's faint song than chattering
Of noisy daws, scattered on clouds of spring.

ERINNA.

I am the tomb of Baucis, the young bride.
Who passes near this tear-besprinkled stone,
"Thou art a jealous god, O Death!" makes moan,
When he the fair memorials hath espied
And they have shown him Baucis' cruel fate,
How those same torches lit the maiden's bier
Which unto Hymen first enkindled were,
And how the songs sung for her bridal state
Sank into wailing cries for one so dear! *

* See Leonidas, Book VII., Epigram 13, p. 23.

19

LEONIDAS.

This is the little farm of Cliton; his
These narrow furrows for the sowing are;
This little wood for cutting twigs is his,
And his this somewhat scanty vine. Ah, well!
Here Cliton passed his four times twenty years.

Dust-loving mouse, go, scamper from my cot!
  The meagre pantry of Leonidas,
Contenting him, for thee sufficeth not.
  Two rolls with salt, such is the fare he has,
Nor asks he better than his father's lot.
What seekest thou then here, thou dainty mouse?
  Thou would'st despise the food whereon I dined.
So hurry off; go try my neighbor's house,
  For here is naught; there thou'lt abundance find!

BOOK VI., EPIGRAM 329.

One, crystal; and one silver brings,
　One, topazes of cost,
For thy birthday fit offerings
　These jewels rich they boast.

But, Agrippina, take from me
　Two verses that I write.
A humble gift I bring to thee
　That envy cannot spite.

BOOK VI., EPIGRAM 355.

Reject not, Bacchus, this poor offering
    A needy mother's ignorant hands have made.
This unskilled picture of my boy I bring,
    Of my Mikythos, asking for thine aid.
    Ah! prosper him, nor let me be afraid!
Though beggarly the gift, despise not thou
My starving poverty, and hear my vow!

LEONIDAS, *or* MELEAGER.

EPITAPH ON ERINNA.

The maiden! The young singer! Like a bee
Stealing the sweets the Muses' flowers among,
Erinna! All too truly hast thou sung:
"Thou art a jealous god, O Death!" * Didst thou
       foresee
How soon thou wert the bride of death to be?

---

* See Erinna, Book VII., Epigram 712, p. 19.

LEONIDAS.

BOOK VII., EPIGRAM 198.

EPITAPH ON A PET LOCUST.

What if small, O passer-by!
  Be this stone! 'tis mine you see.
What if it you scarce descry!
  Philænida gave it me.

Praise her that she held me dear,
  Me, her little locust, singing,
Whether in the stubble here
  Or amid the bushes winging.

Two long years she loved me well,
  Loved my drowsy lullaby;
Me e'en dead did not repel,
  As these verses testify.

24

LEONIDAS, *or* ANTIPATER.

BOOK VII., EPIGRAM 316.

Utter no words, but pass me by
   In silence, nor ask who I be;
Nor seek to know whose son was I.
   E'en silently approach not me,
Go far around and come not nigh!

LEONIDAS.

BOOK VII., EPIGRAM 731.

As hangs the vine upon the garden wall,
On my dry staff I hang, and hear Death call:
"Be not deaf, Gorges, for what boots it thee,
Three or four summers, basking lazily
In sunshine here to lengthen out life's span?"
Speaking these simple words, the old, old man
Betook him to the last home of us all.

BOOK X., EPIGRAM 1.

'Tis time to sail! The chattering swallow's here,
'Neath Zephyr's touch the fields are blossoming;
The boiling waves have smooth'd themselves once
    more,
Rough winds are turned to breezes soft of spring.
Set sail! Set sail! Up anchors and away!
For I, Priapus, sailors all command
That they set sail with goods for every land.

LEONIDAS OF ALEXANDRIA.

Daimon of Argos, in this tomb now lying,

Was he the brother of Deceoteles?

Of Deceoteles.   Did echo, sighing,

Repeat these words? or words of truth are these?

Swift comes the answer: Words of truth are these.

MOERO OF BYZANTIUM.

On Aphrodite's golden porch ye lie,
  O juicy clustering grapes ! The parent vine
No more shall shelter you with honeyed leaf,
  Nor evermore sweet tendrils round you twine.

ZONAS OF SARDIS.

Sand from the water's edge on thy cold head
   And on thine icy body I will heap,
   For on thy grave thy mother may not weep,
Nor see thy ocean-beaten body dead.
The lonely and inhospitable shore
   Of the Ægean caught thee from the wave.
   Receive my many tears and this poor grave,
For thou shalt sail these perilous seas no more!

BOOK XI., EPIGRAM 43.

Give me the cup wrought from the self-same clay
Which bore me and shall cover me some day.

THUCYDIDES.

The great Euripides has for his tomb
All Hellas, though the Macedonian earth
Contains his ashes, since Death found him there.
Hellas of Hellas, Athens was his home,
Thence came the verses that have charmed all
  hearts
And have won every mouth to sing his praise.

RUFINUS.

Where is Praxiteles ? and where, oh ! where
  The hand of Polycleitos who could give
  Such grace to marble as should make it live ?
Who now shall carve Melite's lovely hair,
Her glowing eyes, her neck so dazzling fair ?
  Gone are the artists, gone the sculptors now
  Such beauty with fit temple to endow,
As if an image of the gods it were.

BOOK V., EPIGRAM 48.

Brighter than gold thine eyes, thy cheeks are
  clearer
  Than purest crystal is ; thy mouth so sweet
Is like the reddest rose, but only dearer.
  Thy breast is marble, and thy snowy feet
Like silvery Thetis' are ; and if thy hair
Shows too some silver in its threads, why should
  I care?

BOOK V., EPIGRAM 66.

Meeting by chance my Prodice alone,
I clasp her lovely knees and her implore:
" Save one whose life through thee is nearly o'er,
And sweeten what remains, beloved one ! "
Down fall her tears on hearing what I say,
But gently her soft hands push me away.

BOOK V., EPIGRAM 70.

Thine is all Cypris' loveliness,
   Persuasion's lips are thine;
Thy body is the blossoming
   Of the spring hours divine.

Thy tones are like Calliope's,
   Themis herself so bore,
Thy hands Athene's are, my love,
   Thou mak'st the Graces four.

Book V., Epigram 74.

O Rhodocleia! this wreath I send to thee,
Which I myself of fairest blossoms wove:
A lily, rosebud, an anemone,
And a narcissus with the dew still wet,
A deeply-tinted purple violet;
And crowned with these less proud shalt thou be
      made;
Though fair as they, like them thou too must fade.

BOOK V., EPIGRAM 92.

In all her beauty, proud is Rhodope;
    With brows uplifted she
Scornful returns my greeting when we meet.
    Angry, 'neath haughty feet
She treads the wreaths I hang above her door.*
    Come, wrinkles, and, yet more,
Come, pitiless old age! Come, hasten ye,
    Come bend proud Rhodope.

BOOK V., EPIGRAM 284.

All, all of you I love, save your too kindly eyes,
That something find to love in men whom I despise.

* It was the custom of Greek lovers to hang wreaths above
their mistresses' door.

ERYCIUS.

No more upon thy flute, Therimachus,
Beside the lofty plain thy shepherd's song
Thou'lt tune!   Thy hornèd herds will hear no more
Sweet reedy melodies, while 'neath the shade
Of the broad oak thou liest.   For thou art gone!
Slain by the deadly whirlwind's thunder-blast,
And homeward late the hurrying kine return,
Harassed upon their path by driving sleet.

BOOK VII., EPIGRAM 230.

When, a deserter from the bloody field,

Thy weapons thrown behind, Demetrius,

Thou to thy mother didst return, herself

She pierced thy heart with murderous steel and
  cried:

"Die! that no shame upon thy country fall!

Mine be the fault, not Sparta's, if I have

Suckled a shameless coward at my breast."

SIMMIAS OF THEBES.

BOOK VII., EPIGRAM 22.

Quietly, o'er the tomb of Sophocles,
   Quietly, ivy, creep with tendrils green;
And roses, ope your petals everywhere,
   While dewy shoots of grapevine peep between,
Upon the wise and honeyed poet's grave
Whom Muse and Grace their richest treasures gave.

SIMMIAS.

BOOK VII., EPIGRAM 203.

TO A DEAD PARTRIDGE.

No more from deepest thicket floats
   Thy call, to lead thy mottled comrades on,
Bird of the woods! no more thy flute-like notes
   To shadiest paths invite ; for thou art gone
   Thyself along the path to Acheron.

ANONYMOUS.

BOOK V., EPIGRAM 11.

Cytherea, you who cherish
  Those in jeopardy by sea,
If I, wrecked on dry land, perish,
  Goddess, will you not save me?

BOOK V., EPIGRAM 84.

Would I a rose might be,
  In faintest crimson dressed,
That you might gather me
  And place on your white breast.

BOOK V., EPIGRAM 142.

Does the rose crown Dionysius,
  Or Dionysius crown the rose?
Ah yes! The wearer crowns the crown,
  Which but his beauty shows.

LEONTIUS SCHOLASTICUS.

When Orpheus died, although of him bereft,
  Music lived still.   But Plato,* diest thou,
The Lyre dies with thy dying, naught is left
  Of the old lays that in thy heart and hands found
    life till now.

* Plato, a since forgotten musician.

ANTIPATER.

ON A STATUE OF APHRODITE BY THE SEASHORE.

Small indeed is this my home,
Here where dashes the white foam
      On the shore.
But I love it and rejoice
In the distant threatening voice
      Of ocean's roar.
Sailors, too, for help at sea
Or in love here come to me
      And implore.

Book IX., Epigram 151.

Where is thy beauty gone,
  Thy far-famed beauty, Doric Corinth ? Where
Thy ancient splendor and thy crown
  Of towers that rose in air ?

Thy matrons of the race
  Of Sisyphus, nor countless throngs I see,
For now, unhappy one, no trace
  Is left of them and thee.

War has laid bare the spot
  Whence palaces and temples all are gone ;
And we, the Nereids, whom death touches not,
  We weep thee here alone.

ANTIPATER OF SIDON.

No more, rocks, trees, and savage beasts subdued
  Shall bend, O Orpheus! 'neath thy gentle spell,
Swayed to thy will against their customs rude.
  The raging winds, the roaring ocean's swell,
The swirling snow and hail shalt thou no more
Rule with thy voice's music.   Thou art dead,
Wept by the daughters of Mnemosyne!
And, fellow mortals, how dare we deplore
  The death of earthly children when we see
The bitterest tears Calliope doth shed
  Cannot avail her son from death to free!

Book VII., Epigram 14.

Land of Æolia, Sappho dost thou hide,
That mortal singer who with Muses sang,
And who, by Cypris and by Eros bred,
With Peitho wove the never-dying crown
Of the Pierides, the joy of Greece,
That made thy glory too?   O fateful three !
Who on your shuttles weave the web of life,
Why wove ye not a life perdurable
For her who gave undying gifts of song?

Book VII., Epigram 161.

A. Bird, messenger of Zeus, great Chronos' son,
   How do you dare to perch, so proud, upon
   The tomb of glorious Aristomenes?

B. Because I would all other mortals tell
   He them surpassed as I do birds excel,
   And he was king of those as I of these!
   The timid doves to cowards' graves may cling;
   This hero's courage proudly will I sing!

Book VII., Epigram 303.

Cleodorus, the baby yet unweaned,
   Strove with his tiny feet the deck to tread,
When, 'neath fierce Boreas' blast the ship careened
   And ocean's waves his life extinguishèd.
Ino! you pitied not this little one
As young as Melicertes, your own son!

BOOK VII., EPIGRAM 367.

* * * Egerius compassionate !
For he is dead; and pity too his bride !
On eyes that still sought hers a dark cloud fell,
Extinguishing their light, and life as well.
The envious torch let fade, as fades his breath,
Though lit by sorrowing Hymen and exultant Death.

BOOK IX., EPIGRAM 231.

Round my dry stalk the circling tendrils twine,
    And with another's leaves I'm budding seen.
Once did my spreading branches shield this vine,
    Its roots protecting, when I too was green.
Choose such a mistress who, when you are dead,
    Shall thus repay the love you lavishèd.

AGATHIAS.

All thro' the night I weep,
Till comes the soothing dawn,
Granting my weariness a little rest;
But swallows twittering
Again awake to tears,
Rhodanthe fills with sorrowing thoughts my breast.

Your envious chattering,
O Birds, I pray you cease;
I did not steal the tongue of Philomel.
Go weep for Italus
Among the rocks and caves,
And 'mid the mountains the sad story tell.

On the wild hoopoe's nest
Go find a stormy perch,
And leave me thus to sleep a little space.
And then perchance in dreams
Rhodanthe I may see,
And she may hold me in her arms' embrace.

BOOK V., EPIGRAM 261.*

I love not wine, but shouldst thou wish
    That I its slave might be
Thou needest but to taste the cup
    Then hand it back to me.

Wine that thy lips have lightly touched
    The steadiest head would turn.
And yet from such sweet cup-bearer
    The wine how could I spurn?

For unto me that cup would bring
    From thy dear lips a kiss,
And while I drank would softly tell
    How it received such bliss.

---

* It was upon this epigram that Ben Jonson built his "Drink
to me only with thine eyes."

BOOK V., EPIGRAM 292.*

TO PAUL THE SILENTIARY.

Here the green meadow and the blossoming bough
   Show all the beauty of the fruitful year;
'Neath shady cypresses are singing now
   Bird mothers, brooding o'er their nestlings dear;
From thickets rough the gentle turtles coo,
   And goldfinches out-twitter loud and clear.
But where can I find joy apart from you
   And that shy maiden whom I hold most dear?
A double love home calls me from these shores.
Alas! Law keeps me from her side and yours.

---

* Written by Agathias to Paul the Silentiary when absent on law business. **For the answer, see p. 134.**

BOOK VII., EPIGRAM 204.

ON A DEAD PARTRIDGE.

Exiled from thy rocks and bushes,
   Luckless Partridge, now no more
In thy slender cage of willow
   Shalt thou flutter as before.

Nor shalt warm thy wings vibrating
   In the glowing light of morn,
For thy head from feathered shoulders
   By the cruel cat was torn.

From her gluttony to save thee
   Though I seized thee all too late,
Yet these dear remains I rescued
   From her maw insatiate.

Not too lightly, earth, I pray thee,
   Lie upon my slaughtered pet,
Or, with greedy claws a-scratching,
   His remains she still may get!

Book VII., Epigram 220.

Lais! A wayside tomb showed me this name
 As I along the road to Corinth passed.
 I weeping said: "Ah woman! thou who wast
The torture of young hearts, although by fame
 Alone I knew thee, now my tears thou hast!
 Since thy rare beauty 'neath the earth hath passed,
And Lethe to thy loveliness lays claim."

Book VII., Epigram 569.

Go tell my husband, prithee, passer-by,
 If to my country, Thessaly, thou come,
 That I am dead and that upon the shore
Of Bosphorus within my grave I lie;
 And beg him that he build there at my home,
 To keep me in his mind forevermore,
To me, his wedded wife, an empty tomb.

Book VII., Epigram 602.

Eustathias, still thou art beautiful,
　　But thou art wax, I see ;
No more sweet words on thy lips dwell
　　Whose roses faded be.

Alas! alas! mere dust of earth
　　Ere fifteen years are gone,
And what availed thy father's wealth,
　　And what thy grandsire's throne?

Whoever sees thine image now
　　Must cruel Death upbraid
That 'neath his touch such brilliant glow
　　Of loveliness should fade.

BOOK IX., EPIGRAM 153.

TO TROY.

O city! where are now those walls of thine?
Thy rich and splendid temples, where are they,
With herds of bulls * that to the gods were slain?
The alabaster urns of Cypris where?
And where her tunic woven all of gold?
Where is thy own Athene's image now?
Ah! mighty fate, and war, and flowing time
Have stript thee of them all, and changed thy lot.
To such an end has envious Fate thee brought!
But thy great name no power of hers can hide,
Thy glory shall live on for evermore.

---

* In the ancient temples the skulls of bulls and sheep slain in sacrifice were hung upon the walls; and in some of the temples the cornices were adorned with carved images of the heads of sheep and bulls between the pillars.

ANONYMOUS.

Book VII., Epigram 340.

Nicopolis was laid within this stone
   By Marathonis, whose tears fell like rain
   Upon its marble lid, but all in vain.
For what but sorrow for a man alone
Upon the earth is left, his wife being gone.

Book VII., Epigram 676.

I, Epictetus, was a slave while here,
Deformed in body, and, like Iros, poor,
Yet to the Gods immortal I was dear.

BOOK VII., EPIGRAM 717.

Ye Naiads, and ye frozen pastures, tell
  This tale unto the bees that o'er you wing
  Their wandering way to the far land of spring.
Show what Leucippus, the old man, befell
When he went forth to catch the nimble hare
One winter night.   How death did him ensnare.
  And shepherds now, in many a rustic dell,
  Mourn him who did in neighboring mountains
    dwell,
And swarming bees now miss his fostering care.

ANTIPATER OF THESSALONICA.

Wild beasts has Orpheus tamed with song, and thou
    Orpheus himself hast tamed.
Phœbus has outsung Marsyas, but now
    Thou hast the victory claimed.
Beauty and art are named in naming thee;
    Athene had not thrown
Her pipes aside, if she had known to make
    Such varied melody.
E'en Sleep, in Parithea's arms, would wake
    To hear thy magic tone.

ANTISTIUS.

Cleodemus, a little dancing boy,
The little dancing chorus leads with joy,
   A skin of spotted fawn about him bound ;
The ivy waves above his yellow head.
O Phœbus ! grant that he, full-grown, may lead
   The troops of youths who dance the Bacchic
   round.

ALPHEUS OF MITYLENE.

BOOK IX., EPIGRAM 526.

The mighty throne of the heavens
    Guard, O Zeus, I pray,
For the earth and the ocean tremble
    Beneath the Roman sway.

The unwearied doors of the highest
    Close, I pray, O God!
For the road that leads to Olympus
    Is the only road untrod.

BOOK XVI. [APP. PLAN.] EPIGRAM 212.

ON A SLEEPING EROS.

The flaming torch from thy hands
   I will snatch, O Love! and tear
From thy shoulders the quiver's bands,
   If indeed thou sleepest there.
And may we then enjoy
   From thy arrows a respite brief!
Ah, no! in his dangerous dreams this boy
   Is weaving me some fresh grief.

ZENODOTUS, *or* RHIANUS.

O arid Earth, produce rank briars vying
  With sharpest brambles twisting all around,
That e'en the bird in air above me flying
  Dare press no lightest footprint on the ground
Beneath which I, the misanthrope, am lying,
  Timon, in brother's love to no one bound,
  Who e'en from Pluto never welcome found.

JULIAN OF EGYPT.

BOOK V., EPIGRAM 298.

Charming Maria plays the haughty now.
Come, Justice, ever dear, deal punishment
Unto her saucy pride appropriate!
I ask not, Queen, that death to her be sent,
But that she live until the wrinkled brow
And loose cheek of old age shall be her fate.
My tears avengèd by her snowy hair,
Her beauty, by its loss, shall expiate
The sorrows once it caused when she was fair.

Book VI., Epigram 18.

WITH VOTIVE OFFERING OF A MIRROR.

Lais, her charms by touch of time grown sere,
    Hates her old age and wrinkles to confess,
And bitterly her mirror offers here
    Unto the queen of her lost loveliness:
" Receive this disc,* that to my youth was dear,
    Since that thy beauty fears not time's impress."

Book VI., Epigram 19.

ALSO, WITH VOTIVE OFFERING OF A MIRROR.

Beauty, O Cypris! thou gavest me in vain,
    Since creeping time the victory hath won
    O'er thy most gracious gift ; and, now 'tis gone,
Take then this witness of its loss again.

* The Greeks used polished steel discs for mirrors.

BOOK VII., EPIGRAM 565.

Theodata's self the artist caught.   But yet
Would he had failed and helped us to forget!

BOOK VII., EPIGRAM 576.

O Pyrrho, art thou dead?
    I do not know.
Now Fate's last word is said
    Still doubtest thou?
Here, where thou liest dead,
    Die thy doubts now?

BOOK VII., EPIGRAM 580.

ON A MURDERED MAN.

Oh! never can'st thou dig for me
    A grave so deep that where I lie,
'Neath earth's foundations though it be,
    I shall be hid from Diké's* eye!

* Diké, the Goddess of Justice.

Thou grant'st a grave to me whom thine own hand
    hath slain.
Ah well! May'st thou the same from Heav'n obtain.

BOOK VII., EPIGRAM 582.

Fare thee well, thou drownèd one,
Thou to Hades' shores art gone.
Blame not the waves, but blame the wind,
Since it caused thy death unkind!
But blame thou not the gentle waves
That bore thee to thy fathers' graves!

BOOK VII., EPIGRAM 587.

ON PAMPHILUS THE PHILOSOPHER.

Earth brought thee forth and ocean was thy tomb,
  Ere yet thou shared'st the mansions of the blest:
For a brief space was Pluto's house thy home.
Not conquered by the waves thou sank to rest,
O Pamphilus, but would'st thy glory shed
O'er *all* the dwellings of the undying dead.

BOOK VII., EPIGRAM 590.

A. Illustrious Johannes! B. Call him mortal.
A. And wedded to the daughter of a queen.
B. But mortal none the less. A. Flower of the race
Of Anastasius. B. who yet himself
Was mortal. A. Blameless did he live. B. Ah now
At last of deathless things thou speak'st,
For virtue is victorious over death!

BOOK VII., EPIGRAM 591.

Myself Hypatius'* tomb I dare to call
But claim not that I hold in this small space
That mighty bulwark of the Ausonian race.
His greatness to entomb earth was too small
And gave him to the ocean's vast embrace.

BOOK VII., EPIGRAM 594.

O Theodorus! thy true monument
Upon thy tombstone we should seek in vain,
But find it in the pages where thou hast made
The singers of the past to live again,
Saved from oblivion but by thine aid.

---

* Hypatius, the nephew of Emperor Anastasius, whom the
people crowned against his own wish, was put to death by order
of Justinian, and then cast into the ocean. The Emperor after-
ward raised a tomb to him.

Book VII., Epigram 597.

TO CALLIOPE.

Silent she lies, and hush'd is her sweet song,
    The richest ever heard from maiden's throat.
Strong was her voice, but Moira was more strong,
    No more shall music from her sweet lips float.

Book VII., Epigram 599.

TO KALÉ.

Oh! beautiful by name, and still more fair
In soul than face, she died! With her is gone
The springtide of the graces. She did wear
All Cytherea's grace for him alone
Who was her husband; armèd Pallas she
To other men. When Death took her, what stone
But wept, her from her spouse thus snatch'd to see!

Book VII., Epigram 601.

Alas! for thy sweet spring of joys to come,
Joys numberless, all withered by the blast
From the cold shades of th' all-devouring tomb
That snatched thee from the splendor of the day
Ere yet thy fatal fifteenth year was past!
Now cruel grief with darkness hath o'ercast
Thy spouse and father, tearing thee away
Whose sun of heaven thou, Anastasia, wast.

Book IX., Epigram 654.

This house in poverty's protection lies,
That guardian stern, whom boldest robber flies.

Book IX., Epigram 661.

A happy, happy tree in the wild wood,
This was I once, by winds of heaven caressed,
The haunt of singing birds, as then I stood
Until the woodman's axe had laid me low.
Yet me hath Fate with greater pleasures blest;
The song-birds cling no longer to my bough,
But Craterus himself on me doth rest,
The music of his speech 'round me doth flow.

Book XVI. [App. Plan.], Epigram 130.

The true presentment of sad Niobe
Here weeping for her children you may see.
All this the sculptor gave, but failed alone
In giving life to her the gods made stone.

BOOK XVI. [APP. PLAN.], EPIGRAM 203.

ON A LOVE BY PRAXITELES.

Beneath my feet Praxiteles
   Hath bowed his haughty head,
And then with eager hands has caught
   And me his captive made.

For I, Love, in his heart lay hid
   Till he in metal cast,
And gave me then as pledge of love,
   To Phryne fair at last.

And she again to Eros gave,
   For what could fitter be
As gift of lovers unto Love
   Than Love so fair to see?

BOOK XVI. [APP. PLAN.], EPIGRAM 388.

I twined a wreath of flowers one day,
And lo ! Love 'mid the roses lay.
I seized him by his wings straightway
And plunged him in my wine.
I drank and never more find rest,
But feel love's tremors in my breast.

ANONYMOUS.

O Lacedæmon! thou unconquered one,
   Who inaccessible wert held of old,
Achaian smoke hangs o'er Eurotus' crown,
   And wolves, not sheep, are heard within the fold.
Where once was shade of trees is stript and bare,
And birds with their lamenting fill the air.

BOOK VII., EPIGRAM 737.

Oh! twice unhappy! Here in my last sleep,
By brigands slain, I lie, with none to weep.

BOOK XI., EPIGRAM 53.

The rose but blossoms for a space:
Would'st look for it when past?
Of rose thou'lt find no smallest trace
Save but a thorn at last.

BOOK XVI. [APP. PLAN.], EPIGRAM 129.

ON A NIOBE BY PRAXITELES.

The gods from woman turned me into stone,
Stone to make woman has the sculptor known.

PHILIP OF THESSALONICA.

BOOK VII., EPIGRAM 385.

Hero, Protesilaus, it was thou
Who first taught Troy to dread the Grecian spear,
And the tall trees around thy tomb that rear
Their lofty crown thy wrath with Ilion know,
Since each, when grown so tall that it perceives
Far Ilion's towers, all withered, sheds its leaves.
How hot that hate from whose dead ashes' glow
These trees still draw such hatred of the foe!

BOOK VII., EPIGRAM 554.

This tomb Architeles the sculptor rears
With piteous hands to Agathanor dead;
Yet not by steel was this stone chiselled,
But worn by dropping of a father's tears.
Ah! stone! rest lightly that the dead may say
Truly my father's hand this stone did lay.

BOOK IX., EPIGRAM 575.

The stars shall fade upon the sky,
Or by the sky extinguished be,
The sun shall shine throughout the night,
The thirsty sailor from the sea
Shall drink fresh water, those that die
Shall greet once more the world of light
Before shall be forgot the name
Of Homer, or his verses' fame.

JULIUS POLYÆNUS.

Though thine ear be vexed alway
By the fear of hosts that pray,
And the gratitude of those
Who in prayer have eased their woes,
Yet may we find also grace.
Genius of this holy place,
Zeus of Scheria! hear and nod
Promise of no lying god:
Only let my wanderings cease,
Let long labors end in peace,
In my mother land of Greece.

PAMPHILUS.

No more on fresh green twigs thou'lt sit a-swinging,
No more with sweet and penetrative strain,
Noisy Cicada, shall we hear thee singing,
For a child's hand hath caught thee and hath slain.

ANONYMOUS.

BOOK VII., EPIGRAM 346.

O good Sabinus ! what though small this stone,
  Great was the love that raised it unto thee.
I shall lament thee ever ! Do not thou
  Of Lethe's waters drink one drop for me !

BOOK VII., EPIGRAM 483.

Inexorable Orcus, pitiless,
  The child Callæschrus thou didst tear from life !
  A plaything in the household of thy wife,
His place at home is filled with wretchedness.

Book VII., Epigram 558.

TO RUFINUS.

Hades the blossom of my youth hath gathered
 And hidden it 'neath this ancestral stone.
In vain my birth, although of a good mother
 And of Etherius was I the son,
For thus forbid to reap the fruits of learning
 I languish on the shores of Acheron.
O passer-by! since yet among the living
 Parent or child, thou must be either one,
Therefore lament, this record when thou readest,
 For all my youth and learning so soon gone.

LUCILLIUS.

The miser cries : " Ah, dearest mouse !
Prithee, what dost thou in my house ? "
Mouse answers : " Fear not for thy hoard,
Here I but lodging seek, not board."

81

ANTIPHILUS.

Oak-tree, that stretchest wide thy lofty boughs,
Thy height is goodly shade for men who flee
Immoderate heat.   Thy leafy greenery
Than roof is closer ; house of doves and house
Of crickets.   Me, too, in your rest receive,
O noontide branches, as I lie and drowse
Beneath your tresses, the sun's fugitive.

MARCUS ARGENTARIUS.

This is not love, the eager wish to own
    A woman formed of perfect loveliness,
    'Tis but an eye for beauty to possess.
But when one loves where beauty is unknown,
And burns with madd'ning flames for her alone
    Who in her outward show has ugliness,
    This is indeed love's flame! this, tenderness!
Beauty charms all; not so is true love won.

BOOK V., EPIGRAM 118.

Sweet-breathed Isias, sweetest one, arise,
    And from my loving hands receive this wreath
Now dewy fresh, but fading e'er the skies
    Are red with dawn; here see thy youth and death.

Book IX., Epigram 87.

No longer warble on the oak-tree now,
　It is thine enemy; no longer sing,
O blackbird, sitting on the topmost bough,
　But hasten where the vine climbs clustering
In silvery shade; there may'st thou rest and shrill
　Thy music round her.　Baleful mistletoe
She bears not; but grape clusters grow at will
　On her who is the singer's friend not foe.

Book IX., Epigram 161.

I turn the book of Hesiod in my hands
When suddenly before me Pyrrha stands,
My book upon the ground let fall, I cry,
Why do you bother me, old Hesiod, why?

BOOK X., EPIGRAM 4.

Unloose your cables!   Be your swift sails spread
   All ready, sailors, now to plough the sea!
From smiling zephyr's touch the winter's fled,
   While the blue waves it smooths caressingly.
The chirping swallow builds of straw and clay
   A nest to hold the little nestlings dear;
Fresh blossoms pierce the earth.   Away! away!
   Priapus bids you sail, nor dally here!

ION.

Euripides, all hail !

Who in thy dark-leaved vale

Pierian sleep'st through eternal night.

Though 'neath the dust thou'rt lying,

Yet glory never-dying

Round thee, like Homer, shines forever bright.

DIOTIMUS, *or* LEONIDAS.

At nightfall, driven by snow the hill down sweeping,
  The herds come home alone, for shelter fain.
Alas! alas! Therimachus, thou art sleeping
  Beneath the oak, no more to wake again,
  Since by the fires of Heav'n thou hast been slain.

DIOTIMUS.

Less fierce the terror that the lion wields
    Amid his mountains than Crinagoras,
The son of Micon, 'mid the clashing shields.
    Though small his tomb, salute it as you pass
Small was his country, but was famed to bear
Men who in battle ever valiant were.

BOOK VII., EPIGRAM 475.

Through the wide portals, Scylla, whom we're
    weeping,
      Followed her husband's bier,
On him, Evagoras, with lamentation,
      Prop of his home! she called.
Nor yet again unto her father's dwelling
      Returned the unhappy one,
But died or ever the third month was ended;
      A broken heart her death,
And this sad tomb stands by the dusty roadside
      In memory of their love.

NOSSIS.

Nought sweeter is than love. Whom that doth
    bless
        Regardeth all things less.
If thou first taste of love, then shalt thou see
        Honey shall bitter be!
What roses are, they never know who miss
        Fair Cytherea's kiss.

BOOK VI., EPIGRAM 353.

TO A PORTRAIT OF HIS DAUGHTER MELINNA.

Melinna's very self looks at me here
    With her own gentle face as if she smiled,
How like in all things to her mother dear!
    'Tis sweet to see the mother in the child.

ARCHIAS OF MITYLENE.

Book VII., Epigram 696.

From a shaggy pine-tree thou art swinging
  Thy wild beast's carcass by the tempests tossed ;
Swinging, thing accursed, for beginning
    That monstrous strife with Phœbus, when thou
    wast
On the Celænian promontory dwelling,
  Ah! satyr, we, the nymphs, shall nevermore
Hear the sweet echoes of thy piping swelling
  Among the Phrygian mountains as of yore.

DIOSCORIDES.

BOOK VII., EPIGRAM 229.

Home to Pitana on his shield they bore
   Young Thrasybulus, killed by Argive spears;
Seven open wounds, but all in front, he wore.
   His bleeding son Tynichus, without tears,
Placed on the pyre, then lit the torch, and said :
Mine wast thou and a Greek ! be tears for cowards
   shed !

BOOK VII., EPIGRAM 434.

Eight sons sent Demenete forth to fight
   Against her country's foes ; and on one bier
And in one tomb the mother laid all eight ;
   Then of her loss she said without a tear,
" I bore them, Sparta, but thy sons they were !"

PHILODEMUS.

"Heliodora must thou shun
Ere love for her is in thee begun!"
Thus warned my soul, for she knows well
Love's pangs and tortures to foretell.

Such were her words, but how can I,
If love pursue, have strength to fly?
For she who boldly love reproves,
Already Heliodora loves.

SIMONIDES.

BOOK VII., EPIGRAM 253.

If to die nobly be the greater part
    Of virtue, we, by Fortune, among all
The chosen are, for hastening to give
    Hellas her liberty in death we fall,
Glad in our glory that shall ever live.

BOOK VII., EPIGRAM 258.

The blossom of their youth long since they lost,
    Eurymedon, upon thy shores and tide,
Whether on land, or in swift ships they tost,
    Vainly against the arrowed Medes they tried
Their lances, fighting till they fell.
This fairest tomb their courage rare shall tell.

Book VII., Epigram 300.

Pythonax and his brother, side by side,
Here lie at rest in the cold grave's embrace,
While yet their lovely youth is unfulfilled.
Wherefore their father, Megaristus, willed
A consecrated stone should in this place
Mark his undying thanks for those who died.

BOOK VII., EPIGRAM 496.

Wind-swept Gerania!   Oh, fatal rock!
Would that by Istes and Tanais you were,
'Mong distant Scythians! and not thus near
To snowy Moluris, nor felt the shock
Of the Scironian sea, whose waves now rock
His frozen corpse whose empty tomb doth tell
The fate that in your ocean him befell.

BOOK VII., EPIGRAM 647.

Gorgo, thine arm about thy mother lay;
    One tender speech, it was the last, was thine;
Weeping thou saidst: "Stay with my father, stay
    And bear him other children, mother mine!
Happier in this than she who dies to-day,
    That they may live to soothe thy life's decline."

Book XVI. [App. Plan.], Epigram 204.

ON A LOVE BY PRAXITELES.

The sculptor modelled from his heart,
Receive me, Phryne, nor fear harm,
I conquer not by fiery dart,
But in my eyes is found love's charm.

DIOSCORIDES OF NICOPOLIS.

A slave am I, Timanthes, yes, a slave,

　But as I was thy foster father here

Thou laid'st me, master, in a freeman's grave.

　May'st thou live happy yet for many a year.

And when thou dost rejoin me, master mine,

E'en in the house of Pluto I am thine.

MNASALCAS.

No more, O Locust ! in the fertile furrow,
No more thy clear-toned wings make melody,
Delighting me as in the shade I lie
With tuneful chirping fit to drive off sorrow.

Where foams the sea upon the sunken shore
Let us stand gazing toward the distant grove *
Of sea-born Cypris, and the sacred spring,
Black-poplar shaded, in whose flood the beaks
Of tawny kingfishers dip deep and drink.

* This is supposed to refer to the famous Shrine of Aphrodite at Cnidos.

HERACLIDES.

BOOK VII., EPIGRAM 281.

Spare with thy plough, O Ploughman! spare this
   mound,
Disturb not here the ashes of the dead.
Tears, many tears, upon this spot were shed;
No wheat will flourish on this tear-soaked ground.

CRINAGORAS.

What shall I call you first?   Unhappy one!
What next shall you be called?   Unhappy one!
For you have suffered, but no wrong have done,
Oh! charming woman, who are now no more!
Your face showed forth a perfect loveliness,
And all with perfect love your heart did bless,
And Prote* rightly was the name you bore,
For, sure, such grace was never seen before!

.

* Prote, The First.

BOOK VI., EPIGRAM 253.

O many-watered caverns of the nymphs!
Where coolness trickles from the o'er-hanging rock,
The echoing shrines of Pan with pine-trees crowned,
The lurking valleys hid beneath the cliff,
Or trunks of junipers, decayed and old,
But sacred still to hunters; heaps of rocks,
The piled up shrines of Hermes, will not ye
Receive propitious at Sosander's hands,
The first-fruits of his ever-favored chase?

Book VI., Epigram 345.

Roses of old oped with the opening year,
    But we our crimson chalices throw wide
In winter, greeting thus thy birthday, near
    To that blest day when thou shalt be a bride.
If us upon thy head thou deign to wear,
    O loveliest woman ! there to be espied
Were than the sun of spring to us more dear !

Book VII., Epigram 633.

The moon, arising on the verge of twilight,
Hath clouded all her beams to hide her tears,
Since that Selene, her most lovely namesake,
Doth life relinquish and to shades descend.
For she would share death's darkness with the
    maiden
Round whom she flung the beauty of her light.

CHÆREMON.

Eubulus, son of Athenagoras,
  Thou wert outstripped by all in length of days,
  But in thy measure of deserved praise,
Indeed none is there who can thee surpass.

Book VII., Epigram 721.

Sparta 'gainst Argos sallies forth to fight;
  Like arms, like numbers each pours o'er the plain;
Thyreæ the prize; in combat they unite,
  Nor ever think to see their homes again.
  The birds shall be sole heralds of the slain.

CAPITO.

Beauty, alone, may please, not captivate;
If lacking grace, 'tis but a hookless bait.

BIANOR OF BITHYNIA.

BOOK VII., EPIGRAM 387.

Ere for Theonoë my tears were dried,
    Though hope awakening clung around our boy,
An envious Fate hath torn him from my side,
    My little son, all that I had of joy!
Hear, Diké, this one prayer from heart oppressed,
And place my child on his dead mother's breast!

BOOK XI., EPIGRAM 364.

This man, despised, a slave, an outcast thing,
Was loved of one and in one soul was king.

ASCLEPIADES.

Book V., Epigram 145.

O wreaths! remain here hanging on this door,
   Nor, hasty, shake your leaves,
Your leaves, that I have drenchèd with my tears,
   Such tears as lovers shed.
But when you see the door softly unclose,
   Let fall your bitter dew
Upon her head, that her light golden hair
   May thus drink in my tears.

Book V., Epigram 189.

Long, very long and wintry is the night,
Already are the Pleiads sinking low,
While up and down I pace before her door
In the fast falling rain in sorry plight,
Charmed by a cruel one, not loving, no!
But pierced by burning grief to my heart's core.

CALLIMACHUS.

They tell me, Heraclitus, thou art dead,
And many are the tears for thee I shed,
With memories of those summer nights opprest
When we together talked the sun to rest.
Alas! my guest, my friend! no more art thou;
Long, long ago wert ashes, and yet now
Thy nightingales* live on, I hear them sing,
E'en death spares them, who spares not anything.

Leontichus, an alien in this land,
  Found thee, poor drownèd sailor, on the shore,
And dug for thee this grave here in the sand;
  And though he wept the while, his tears were more
For his own perilous calling than for thee,
Since he too, like a gull, sweeps o'er the sea.

* "Nightingales" refers to the poems of Heraclitus.

108

ARCHIAS.

From Eros would'st thou flee?
Vain shall the struggle be:
Canst thou escape on foot
One winged to follow thee?

EPITAPH ON A MAGPIE.

I who often did reply
    To fisherman's or shepherd's song
With my merry mocking cry,
Like an echo from the sky,
    Now, without a voice or tongue,
Silent, fall'n to earth, I lie.

Book VII., Epigram 278.

I, Theres, though a dead man, cast away
   And flung up by the waves upon this land,
Where the sea-smitten cliff beside I lay,
   Having found burial from a stranger's hand,
The hateful, the ill-wishing deep too near;
   Never shall I forget that sleepless shore
For still is booming, booming in my ear
   The thud of ocean's waves forevermore;
And I among the shades am wretchedest,
Whom e'en the grave gives not unbroken rest.

MACEDONIUS.

Thy mouth is grace itself; thy cheeks are flowers;
      Thine eyes are love's own fire;
      Thy fingers clasp the lyre;
Our ears thy voice doth charm; these eyes of ours,
      Slaves of thy loveliness,
      Thy beauty doth possess.

APOLLONIDES.

BOOK VII., EPIGRAM 180.

Death's lottery is changed, and in thy place,
O Master, I have filled a gloomy tomb;
When I, thy slave, was digging underground
A tearful grave, to place thy body there,
The earth fell in about me.   Yet not sad
Are Pluto's shades to me; since thou'rt my sun.

ANYTE, *or* LEONIDAS.

Unto the locust, nightingale of fields,
  And the cicada, who was wont to drowse
  Through summer heat amid the oaken boughs,
This common tomb the maiden Myro builds,
And, like a child, weeps that she could not save
These twain, her cherished playthings, from the
  grave.

BOOK VII., EPIGRAM 215.

ON A DEAD DOLPHIN.

Alas! from the ship-laden sea I may
Dash joyful upward through the waves no more,
And splashing 'mong the fair ships' prows shall I
No more delighted with my image play.
For a black tempest drove me 'gainst the shore,
And here on ocean-beaten sands I lie.

BOOK VII., EPIGRAM 538.

He was a slave in life who lieth here,
Now, being dead, he is Darius' peer.

ANTIPHILUS.

BOOK VII., EPIGRAM 399.

E'en in their tombs let them lie separate,

    These sons of Œdipus, not side by side.

Though they are dead, yet living in their hate,

    Nor in one boat would they cross death's dark
    tide.

And on the funeral pyre the flames divide,

By one torch lit, each struggles with his mate.

ALCÆUS OF MITYLENE.

BOOK VII., EPIGRAM 536.*

This dead old man here lying
Upon his tomb produces
No vine with grapes refreshing,
But brambles rough and thorny,
Wild fruits best fit for choking.
The dry-lipped, thirsty traveller
Hipponax' tomb who passes
Shall pray that corpse so kindly
May sleep and know no waking.

---

* Satirical epitaphs like this one are by no means rare in the Anthology, and it was thought well to give one as a specimen.

ÆSCHYLUS.

They fought till death, unheeding the spear's thrust,
And saved their fertile land where cattle fed;
Their glory liveth, though themselves lie dead,
Who made their strenuous stand in Ossa's dust.

Athenian Æschylus, Euphorion's son,
  In his last rest doth 'neath this stone abide,
  'Mid the wheat-fields of Gela, where he died.
Be witness of his manhood, Marathon!
And also let the long-haired Persians tell
His courage, which they knew, and overwell!

ÆSOP.

O Life, what refuge have we fleeing thee,
Save in Death only ?   Infinite, in truth,
Thy sorrows are, and unendurable
As unavoidable.   Doubtless there are
Some beauties and some charms in Nature's gift—
The earth, the stars, the sea, the moon, the sun,
But all the rest is only grief and fear.
And if perchance some happiness be there,
There too is Nemesis who takes revenge.

PALLADAS.

ON A MARBLE EROS.

Stript of his bow and fiery darts
Love thus can smile and harmless be,
These flowers and dolphins that he holds
Show forth his sway on earth and sea.

ANONYMOUS.

Euripides, all Greece thy monument,
Thou art not dumb, indeed, but eloquent.

And thou, Protagoras, thou art, we know,
The shining arrow of philosophy,
But as thy truth straight to our hearts doth go,
Not wounded, but most gently soothed are we.

BOOK VII., EPIGRAM 155.

EPITAPH ON PHILISTION OF NICÆA, AN ACTOR.

He who by wakening laughter much did cheer
Man's sorry lot, Philistion, lies here;
Full oft when living he for dead hath passed,
But now in other fashion's dead at last.

BOOK VII., EPIGRAM 137.

EPITAPH ON HECTOR.

Judge not me, Hector, by this tomb,
Nor measure by this little mound
　　The antagonist of Greece.
The Iliad, Homer, are my grave,
The flying Grecians, Greece itself,
　　All these my monuments.
Though slight the dust above me piled
Not mine the fault ! At hostile hand
　　Of Greeks I burial found.

PLATO.

I, the proud Lais, to whose door once came
Troops of young lovers, and whose toy was Greece,
I consecrate to Cytherea now
My mirror—since I can no longer see
Myself reflected there as once I was
And would not see, alas! as now I am.

We, who had passed uninjured through the swell
Of the deep-voiced Ægean's mighty waves,
On Ecbatana's plain have found our graves—
Eretria, renowned of old, to thee,
To thee, dear native land, our last farewell!
Athens farewell, farewell belovèd sea!

BOOK VII., EPIGRAM 669.

Thou gazest on the stars, my star !
Ah ! would that I might be
Myself those skies with myriad eyes,
That I might gaze on thee.

BOOK IX., EPIGRAM 823.

Hush'd be the leafy rocks of the Dryads !
Hush'd be the streams from those rocks that
 · spring !
Hush'd be the ewes that bleat to their lambkins !
For Pan himself his song would sing ;
His flexible lips to the pipes he presses,
And the water and wood nymphs in dances spring.

BOOK XVI. [APP. PLAN. I.], EPIGRAM 13.

'Neath this tall pine,
That to the zephyr sways and murmurs low,
    Mayst thou recline,
While near thee cooling waters flow. ·
    This flute of mine
Shall pipe the softest song it knows to sing,
And to thy charmèd eyelids sleep shall bring.

BOOK XVI. [APP. PLAN. IV.], EPIGRAM 210.

We came upon a shady grove;
Like crimson apples, hidden there
We found, on roses lying, Love;
Of bow and quiver he was bare,
They hung above him on the tree,
While he lay sunk in slumbers deep;
His dainty lips that smiled in sleep
Were clustered round by tawny bees,
As though in honey they would steep.

PAUL THE SILENTIARY.

How long our loving glances shall we hide?
And fear to meet each other's eyes, how long?
Let our love speak, and if it be denied,
If each to each in love may not belong,
The sword shall be the healer of our pain:
Sweet is or life or death, shared by us twain.

BOOK V., EPIGRAM 230.

With one hair from her head did Doris tie
   My hands, her captive I!—
I laughed aloud, so easy 't seemed at first
   The golden thread to burst,
But when I find not all my strength can tear
   Sweet Doris' single hair
From off my fettered hands, unfortunate
   I groan my bitter fate,
Since evermore this hair the chain shall be
   By which she leadeth me.

BOOK V., EPIGRAM 241.

The moment comes to say to thee "farewell!"
Yet by thy side I linger silently.
Must I then go?   Such parting were to me
More dreadful than the darkest gloom of Hell,
For thou art as my very light of day,
But day is silent, and thy gentle voice
More than a Syren's song makes me rejoice,
And round thy lips all my soul's longings stay.

Book V., Epigram 250.

Sweet are the smiles of Lais! and how sweet
    Tears from her charming eyes!
But yesterday she leaned on me and wept
    Without a cause and moaned.
I kissed her, but her tears still fell like rain.
    " Why weepest thou ? " I prayed.
" I feared lest thou should'st leave me," murmured
    she,
    " For men are never true."

Book V., Epigram 254.

I swore from thee till the twelfth dawn to part,
O fair young girl, but could not keep my vow,
For when the morrow in the sky shone bright
Like twelve long months the hours oppressed my
  heart.
And for your luckless friend beseech the gods,
Lest on their scroll of punishment they write
My broken oath, and, dear one, gracious be
To one who fears the wrathful gods and thee!

Book V., Epigram 256.

Last evening Galatea closed her door,
With scornful words my very face before.
Disdain, they say, kills love. Alas! not so,
Disdain but makes a lover's madness grow.
I swore I would remain a year away,
But suppliant at her door am found to-day!

Book V., Epigram 270.

No crown the rosebud needs, and thou,
Thou need'st no broidered veils and gems to wear;
Gold adds no brightness to thy flowing hair,
Pearls are less white than are thy neck and brow.
From purple depths of the Indian jacinth gleams
A sparkling fire, but thine eyes shine more bright,
Thy fresh lips and thy graceful form that seems
A goddess's could not have greater might
If Cytherea's girdle thou shouldst wear.
To approach such loveliness I should not dare,
Did not thy gentle eyes my heart invite
The sweet hope that I read in them to share.

Book V., Epigram 293.

(An answer to Agathias.*)

True love doth know law cannot separate
A man in love from his self-chosen mate,
And if your task of law from love you wrest,
Impetuous love dwells not within your breast.
What love is this if such small space divide
You from the girl you wish to make your bride?
The power of love oft made Leander brave
For his beloved's sake the midnight wave.
You, friend, have ferryboats! but you adore
Athenè, and ne'er look on Cypris more.
Come, say, doth Love or Law you most rejoice?
No man can serve them both, so take your choice!

* See p. 50.

Book V., Epigram 301.

Though far beyond Meroë thou should'st go,
Love, winged Love shall bring me unto thee:
Or if thou wander to the Orient,
To seek the dawn that is less fair than thee,
My feet should follow thee, however far.
And if a tribute from the sea I bring
To offer thee, young girl, reject it not!
The goddess of the ocean sends it thee :
For she, in grace and charms by thee o'ercome,
No longer claims to be the loveliest!

Book VII., Epigram 307.

My name—why tell it?—Country—matters not—
From famous blood—what if from poor thou come?
Of honorable life—Had'st thou been bad, then
    what?
Here I lie now—who says this, and to whom?

DIOGENES LAERTIUS.

O Socrates !   Now as the guest of Zeus
Thou drinkest; thou, whom gods have called the
    wise,
With godlike wisdom.   Nor didst thou refuse,
When Athens offered thee the hemlock draught,
But, with thy mouth, herself hath poison quaffed.

Book VII., Epigram 129.

A noble wish was thine, nobly pursued,
O Zeno, when the tyrant thou would'st slay,
Hellas to free from her forced servitude.
But ah! thou wert the slain; the conqueror, he,
He crushed thee into dust—what do I say?
Thy body did he crush, but never *thee !*

INDEX OF AUTHORS

www.ingramcontent.com/pod-product-compliance
Lightning Source LLC
Chambersburg PA
CBHW021811190326
41518CB00007B/540